X-506-66-156

APOLLO UNIFIED S-BAND SYSTEM

by
K. E. Peltzer
Manned Flight Support Office

April 1966

GODDARD SPACE FLIGHT CENTER
Greenbelt, Maryland

APOLLO UNIFIED S-BAND SYSTEM

by
K. E. Peltzer

FOREWORD

This paper was originally prepared for and given at the 1965 IEEE International Electronics Symposium.

ABSTRACT

This paper describes the Apollo Unified S-Band (USB) System. The areas covered are USB station locations, uplink and downlink spectrums, ground station and spacecraft gross parameters, general theory of operation of the system, and the functions of both the ground station and spacecraft Unified S-Band System. It is intended for those people who would like a general understanding of the USB System, but who do not intend to work in the detailed engineering phases of the program.

APOLLO UNIFIED S-BAND SYSTEM

by
K. E. Peltzer

Project Apollo is considerably different from the previous manned space projects. The basic mission of both Mercury and Gemini was launch, insertion into earth orbit, orbiting of the spacecraft about the earth a certain number of orbits, re-entry, and recovery. Apollo is considerably different since injection into translunar flight and the capture of the spacecraft by the moon's gravitational field must be considered. The Command Module (CM) - Lunar Excursion Module (LEM) separation and LEM descent to the lunar surface are phases that must be monitored by the earth ground stations. In addition, the LEM ascent and rendezvous with the CM are powered flight phases that need to be closely monitored. Perhaps the most critical phase in the return trip to earth is that of re-entry. During this period, the spacecraft must enter a very narrow corridor in order to be captured by the earth's gravitational field at such an angle as to avoid extreme overheating that might cause excessive melting of the spacecraft. During the "blackout" period, it is desirable to know the spacial coordinates of the spacecraft at all times with respect to the ground station. These phases have caused communications requirements that are significantly different from previous manned space missions.

Projects Mercury and Gemini both utilized a C-band beacon on the spacecraft craft that could be interrogated by a high-duty-rate ground radar as the primary means of tracking. Uplink voice and command data, and downlink voice and telemetry data were sent via UHF and VHF systems. The Apollo telecommunications system is uniquely different since the basic capability exists to send all information to the spacecraft and receive all information from the spacecraft with the Unified S-Band System. There will be other systems on the early Apollo flights such as C-band beacons and VHF telemetry; however, the later 500 series which involves the lunar phase will utilize the Apollo Unified S-Band System for primary tracking of and communications with the spacecraft.

Figure 1 is a summary of the capabilities that exist for both the ground-to-air link and the air-to-ground link with the Unified S-Band System. The ground-to-spacecraft link consists of voice, digital command data, and ranging signals. All of this information can be modulated onto one (1) common S-band carrier as a result of modulating two (2) different subcarriers - one (1) with voice and one (1) with digital commands. The command data and voice subcarriers are combined with the ranging signals to form the modulating waveform for the S-band carrier. This information is sent to the spacecraft via one (1) common S-band carrier. The spacecraft-to-ground link is similar, however, there are more types of information that must be transmitted from the spacecraft. Astronaut voice and telemetry data, the latter originating in the spacecraft from the various sensors either on the astronauts or on-board systems, are two (2) important pieces of information that must be transmitted from the spacecraft. In addition, real time television for the lunar phase, astronaut biomedical data, and ranging signals must be transmitted back to earth. It should be pointed out that the emergency voice and emergency key modes are back-up modes required in the event of a spacecraft equipment failure. These modes would allow the very minimal amount of information to be obtained from the spacecraft if such a failure occurred.

APOLLO UNIFIED S-BAND SYSTEM CAPABILITIES

GROUND TO SPACECRAFT

1. VOICE
2. DIGITAL COMMANDS
3. RANGING SIGNALS

SPACECRAFT TO GROUND

1. VOICE
2. TELEMETRY
3. TELEVISION
4. BIOMEDICAL DATA
5. RANGING
6. EMERGENCY VOICE
7. EMERGENCY KEY

Figure 1. Apollo Unified S-Band System Capabilities

Figure 2 indicates the location of the major sites in the World-Wide Manned Space Flight Network utilizing the Unified S-Band System. The basic Apollo mission is broken down into six (6) phases. These are launch, insertion, near-earth orbit phase, injection, the lunar phase, and re-entry through the earth's atmosphere. Each station has accompanying X's that denote which phases of the mission are covered. A small "s" denotes those stations having a 30-foot parabolic antenna. The "L" denotes those stations utilizing an 85-foot parabola. Three (3) of the Apollo ships utilize a standard 30-foot antenna. The other two (2) ships are utilized for re-entry and the S-band equipment has not been completely specified. It should be pointed out that the three (3) prime 85-foot stations are Goldstone, California; Canberra, Australia; and Madrid, Spain. These three (3) stations are dual stations. A dual station is indicated by a "D" to the left of the site. By dual station it is meant that there is only one (1) antenna at the station, however, dual reception equipment is available for receiving simultaneously from several different spacecraft on different frequencies. The transmitting equipment consits of one (1) transmitter for a dual 30-foot and two (2) transmitters for transmitting two (2) different uplink carriers for the 85-foot sites. The 30-foot antenna stations that are dual are Carnarvon, Australia; Cape Kennedy, Florida; Hawaii; Guam; and Ascension.

UNIFIED S-BAND SITES

SITE	CODE*	LAUNCH	INSERTION	NEAR EARTH	INJECTION	LUNAR	RE-ENTRY
CARNARVON	D,S			X	X		
BERMUDA	S	X	X	X			
TEXAS	S			X			
CAPE KENNEDY	D,S	X	X	X	X		
GUAYMAS	S			X	X		
HAWAII	D,S			X	X		
GUAM	D,S			X	X		
ASCENSION	D,S			X	X		
CANARY ISLANDS†	S			X	X		
ANTIGUA†	S		X	X			
GRAND BAHAMA I.†	S	X	X	X			
GOLDSTONE	D,L					X	
CANBERRA	D,L					X	
MADRID	D,L					X	
SHIP NO. 1	D,S		X		X		
SHIP NO. 2	D,S				X		
SHIP NO. 3	S				X		
SHIP NO. 4							X
SHIP NO. 5							X
AIRCRAFT (8)					X		X

*CODE
 D - Dual Stations
 S - 30 Foot Ground Antenna
 L - 85 Foot Ground Antenna

†Canary Islands, Antigua, and Grand Bahama Islands are proposed sites only.

Figure 2. Unified S-Band Sites

In addition, there are two (2) ships that will be dual 30-foot S-band stations. The insertion ship in the Atlantic Ocean will be a dual station as well as the Indian Ocean ship used for injection. A dual transmitting capability is implied for all 30-foot dual S-band stations, however, one (1) power amplifier amplifies two (2) uplink carriers.

The 85-foot Apollo Manned Space Flight Network (MFSN) stations are utilized primarily for the lunar phase of the mission. This is due to the fact that these antennas have 9 db higher gain than the 30-foot antenna, which is required for some of the primary communications modes during the lunar phase. It is interesting to point out that the 85-foot antenna has a 3 db beamwidth which covers an area at lunar distances approximately 1,400 miles in diameter while the moon's diameter is about 2,160 miles. Thus, the overall 85-foot antenna movement is very small during coverage of the spacecraft at lunar distances. In addition to the MSFN 85-foot stations, it is planned to utilize the Jet Propulsion Laboratory's (JPL) Deep Space Instrumentation Facilities (DISF) which are also 85-foot stations. These can be utilized either as back-up facilities in the event that a prime station has a failure or they could be utilized to cover either the LEM or CM if the separation between vehicles were too large for both to be covered by the 3 db beamwidth of one (1) 85-foot antenna.

The Apollo aircraft will be utilized to provide voice communications relay and telemetry recording during the injection burn and re-entry phases of the mission. The aircraft receives voice from the spacecraft via VHF and the Unified S-Band System; demodulates the voice, and re-transmits the voice to the ground communications network via HF. The aircraft receives voice from the ground, and automatically transmits it to the spacecraft via UHF or Unified S-Band. The S-band system in the aircraft also demodulates the telemetry from the Unified S-Band system, and records the telemetry data for post-flight analysis.

Figure 3 is a summary of the various phases of the mission covered by the Unified S-Band System. During the early 200 series Apollo missions, great reliance will be placed upon the C-band tracking radars, the UHF command system, and the VHF telemetry systems. However, the later 500 series Apollo missions will utilize the Unified S-Band System as the primary tracking and communications system. The only phase of the mission in which the Unified S-Band System does not play a major role is the recovery phase. In this phase, HF recovery systems similar to the recovery system which has been used on both Mercury and Gemini are utilized. In summary, it may be said that the Unified S-Band System on the later 500 series shots will be the primary communications system from lift-off on the pad at Cape Kennedy through re-entry into the earth's atmosphere.

PHASES OF MISSION COVERED BY USB

1. LAUNCH
2. INSERTION INTO EARTH ORBIT
3. EARTH ORBIT
4. INJECTION INTO TRANSLUNAR FLIGHT
5. LUNAR PHASE
6. RE-ENTRY

Figure 3. Phases of Mission Covered by Unified S-Band

Figure 4 indicates the various spacecraft that can be communicated with via the Unified S-Band System. As can be seen, the Unified S-Band System plays the major communications role between the CM, LEM, S-IV B, and the ground stations.

SPACECRAFT COMMUNICATED WITH VIA UNIFIED S-BAND SYSTEM

1. COMMAND MODULE
2. LEM
3. SATURN S-IV B

Figure 4. Spacecraft Communicated with Via the Unified S-Band System

The Unified S-Band System on the Command Module (CM) is slightly more intricate than that aboard either the LEM or the Saturn S-IV B. This is due to the fact that the CM is in a sense the basic home of the astronauts. The astronauts ride in the CM until they get in the vicinity of the moon. At this point, two (2) astronauts transfer from the CM to the LEM at which time the CM and the LEM separate and the LEM makes its descent to the moon. After separation, both spacecraft can be communicated with via the USB System. The S-IV B is separated from the CM/LEM during transposition which occurs about 12,000 miles from earth. Each of these spacecraft have the basic capability to demodulate the ranging code and remodulate it on the downlink carrier for transmission to the ground.

Figure 5 is a very basic diagram of the Unified S-Band ground station. As can be seen, there are two (2) antennas shown on this block diagram associated with one (1) ground station. The smaller antenna, called the acquisition antenna, is used as an acquisition aid. This antenna has a beamwidth considerably larger than the main antenna for both the 85-foot and the 30-foot antennas. As the spacecraft comes over the horizon, the acquisition antenna is used for initial angle acquisitions. Although the acquisition antenna makes initial acquisition more probable the angle information initially is reasonably coarse. However, this pointing information is good enough to position the large antenna on the spacecraft. A diplexer is provided to isolate the transmit and receive frequencies which are separated by about 180 Mc. This isolation is necessary since transmission and reception occur simultaneously. These frequencies are separated by a translation factor of 240/221 that occurs in the transponder plus the two-way Doppler frequency. A large amount of isolation is required due to the fact that the threshold of the main receiver is in the region of -157 dbm for the narrow bandwidth in the PM channel. This diplexer provides approximately 190 db of isolation between the transmit and received frequencies.

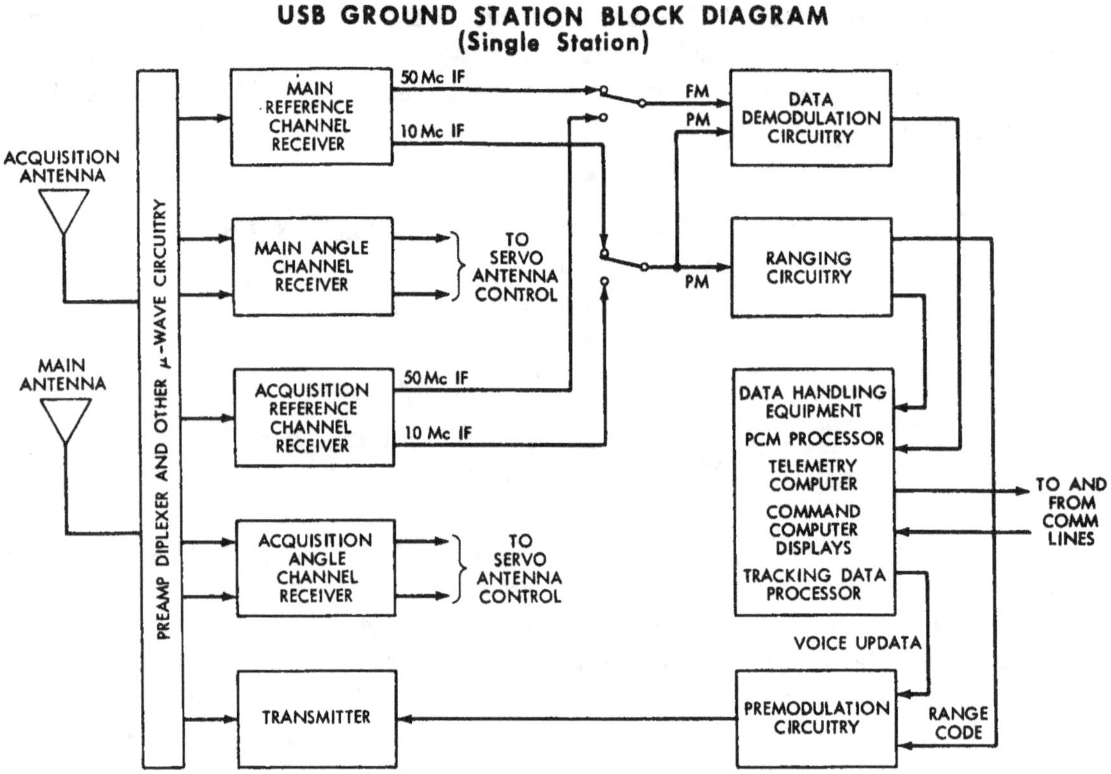

USB GROUND STATION BLOCK DIAGRAM
(Single Station)

Figure 5. Unified S-Band Ground Station Block Diagram

The acquisition reference channel receiver is fed by the sum output of the acquisition feed system. This receiver utilizes a narrow band phase-lock loop in the front end which locks to the received frequency. There are two (2) data outputs associated with the acquisition reference channel receiver. One (1) output is a 50 megacycle frequency modulated output and the other is a 10 megacycle phase modulated output. These two (2) outputs are required to receive the two (2) basic modes that can be transmitted from the spacecraft - one (1) is the FM mode and the other is the PM (phase modulated) mode. Two (2) angle acquisition receives are utilized with the acquisition antenna. These receivers are fed by the X and Y difference outputs of the feed system. Any error developed in the difference pattern is proportional to the angular difference between the electrical axis of the antenna and the angle of the spacecraft. The angle error generates a voltage error in the acquisition receivers that is fed to the servo system. This voltage, being proportional to the angular error, is used in the servo control system to position the antenna so that the electrical axis coincides with the angle of the spacecraft. After the acquisition antenna has developed adequate pointing data, the main antenna is switched in to receive from the spacecraft. The uplink frequency is sent to the spacecraft; the spacecraft receives this frequency; locks to it; and transmits the downlink to the ground station. A two-way coherent lock can then be achieved, upon which time the full two-way communications and tracking can begin.

The main reference channel receiver has a very narrow-band carrier tracking phase lock loop. This loop has three (3) switchable filters resulting in loop bandwidths at threshold of either 50, 200, or 700 cps. This loop locks to the downlink carrier frequency and tracks it as the frequency changes due to the change in the radial component of the spacecraft velocity. The Doppler frequency is derived from the voltage-controlled oscillator in this main carrier tracking phase-lock loop. There are two (2) data outputs from the main reference channel receiver - one (1) output is the 50 megacycle FM output and the other is the 10 megacycle phase modulated output. These outputs feed two (2) sets of data demodulators - one (1) set for the FM data mode and another set for the PM data mode. In addition, the phase modulated output is fed to the ranging receiver. The received ranging code is correlated with the receiver ranging code and a correlation function is developed at a time from which the range to the spacecraft can be derived. As previously mentioned, Doppler information is derived by comparing the received frequency with the transmitted frequency.

The main angle channel receiver is fed by the X and Y difference outputs of the 4 horn feed system. Any angle error develops an error voltage which again is proportional to the angle between the electrical axis of the antenna and the spacecraft angle. This voltage is fed to the servo antenna control which drives the antenna to the position where the electrical axis will coincide with the angle of the spacecraft. In this manner, the antenna is constantly being corrected to keep the beam centered on the spacecraft.

The premodulation circuitry in Figure 5 consists of the circuitry required to generate the uplink spectrum. Thus, part of this is the subcarrier oscillators. One subcarrier is modulated by voice and the other is modulated with up data commands derived from the command computer. The ranging code is also fed to the premodulation circuitry. Here, the modulated subcarriers and ranging code are combined to form the modulating waveform for the S-band carrier. The S-band carrier is derived from a frequency synthesizer which operates at approximately 22 megacycles. The modulated S-band spectrum is then amplified by a 20 kw klystron power amplifier which feeds the antenna via the diplexer.

The outputs of the main reference channel receiver are followed by a carrier tracking phase-lock loop for the 10 Mc PM output and a modulation tracking phase-lock loop for the 50 Mc FM output. A set of demodulators follow these loops for the two (2) different data outputs. The output of the 50 Mc FM carrier demodulator feeds a telemetry subcarrier demodulator, a voice subcarrier demodulator, and a TV monitor. Similar circuitry follows the 10 Mc PM carrier demodulator, except TV is not associated with this mode. However, provisions are made for demodulation of the emergency key, emergency voice, and biomedical data in the PM demodulator channel.

A great amount of data processing equipment follows the signal data demodulators. The output of the telemetry demodulator feeds the PCM processor that decommutates the telemetry bit stream through stored information on the frame synchronization patterns. The decommutated data is fed to the 642 B modified computer where it is stored, processed, and pertinent information extracted for display or transmission to the Mission Control Center (MCC). Another 642 B computer is used for storing the digital commands. Digital commands generated at the MCC are sent to the remote sites and stored in this computer. The commands can then be called up for transmission to the spacecraft via the S-band uplink. In addition, there are a number of displays at the remote sites that are driven from outputs of the telemetry computer.

The Tracking Data Processor (TDP) assimilates all tracking data derived at the site. This includes the outputs of the shaft encoders on the antenna, biased doppler frequency, range, time of day and transmitter VCO frequency. This data is formatted and sent to GSFC and MSC for orbit computations.

Figure 6 outlines the basic characteristics of the ground antennas. The small, primary antenna is a 30-foot parabola with a Cassegrain feed system.

GROUND ANTENNAS

1. 30 FOOT PARABOLA WITH CASSEGRAIN FEED SYSTEM:
   GAIN 43 db
   BEAM WIDTH 1.0°
   3 FOOT ACQUISITION ANTENNA MOUNTED AT APEX OF QUADRIPOD OF 30 FOOT DISH

2. 85 FOOT PARABOLA WITH CASSEGRAIN FEED SYSTEM:
   GAIN 52 db
   BEAM WIDTH 0.35°
   6 FOOT ACQUISITION ANTENNA MOUNTED AT APEX OF QUADRIPOD OF 85 FOOT DISH

Figure 6. Ground Antennas

The Cassegrain feed system is essentially part of the angle measuring system for the USB equipment in that it develops a difference pattern on the signal received at the antenna for both the X and Y axes. This information, as previously mentioned, is used to keep the antenna positioned on the spacecraft during the pass over the station. The gain of the 30-foot antenna at the transmit frequency is approximately 42 db, and about 43 db at the receive frequency. The beamwidth of this antenna is one degree. A 3-foot acquisition antenna is used with the 30-foot antenna and is mounted at the apex of the quadripod. The 85-foot antenna utilized a similar feed system to the 30-foot antenna. This antenna has a gain at the receive frequency of approximately 52 db with an accompanying beamwidth of 0.35 degrees. As shown, the acquisition antenna for the 85-foot dish is a 6-foot parabola. Thus, both acquisition antennas are mounted to be in line with the electrical axis of the larger antenna.

Figure 7 gives an outline of the basic ground system characteristics. The power amplifier utilized in the USB System is a 20 kilowatt CW klystron amplifier. This power amplifier has sufficient bandwidth to amplify simultaneously the LEM's uplink carrier and the CM's uplink carrier. This is required at a dual 30' station. A dual 85-foot station has two (2) complete transmitters in order to provide more gain during those mission phases when it is necessary to have more uplink signal power. Each power amplifier is fed by a separate frequency exciter and synthesizer that generates the respective frequency for either the LEM or CM.

GROUND SYSTEM CHARACTERISTICS

1. POWER AMPLIFIER - KLYSTRON
   POWER OUTPUT - 20 kw cw

2. PREAMPLIFIER - PARAMETRIC AMPLIFIER
   GAIN - 20 db
   NOISE FIGURE - 1.7 db
   BANDWIDTH - 30 Mc

3. CARRIER FREQUENCIES - UPLINK
   2106.4 Mc COMMAND MODULES
   2101.8 Mc LEM OR S-IV B

4. VOICE SUBCARRIER
   FREQUENCY - 30 kc
   FM MODULATION - 100 cps TO 3 kc

5. UPDATA SUBCARRIER
   FREQUENCY - 70 kc
   MODULATION - 1 kc SYNC TONE; 2 kc SINE WAVE BI-PHASE MODULATED AT 200 cps

6. RANGE CODE
   LENGTH OF SHORT CODE - L_1 - 70 MILLISEC
   LENGTH OF LONG CODE - L_2 - 5.4 SECONDS

Figure 7. Ground System Characteristics

The receiver preamplifier is an uncooled parametric amplifier. This preamplifier has a gain of 20 db at the receive frequencies and its gain is reasonable flat over the 30 Mc frequency range of interest. In addition the noise figure as specified for this device is 1.7 db referred to 290° Kelvin.

There are two (2) basic carrier frequencies that can be generated by the USB ground stations. These are 2106.4 Mc for the CM and 2101.8 Mc for the LEM. These frequncies are nominal and can be changed, if necessary, by the synthesizer control. It is important to note that the LEM uplink carrier frequency is also used for the S-IV B. In order to maintain a coherent two-way link, it is desirable to establish a two-way link with only one of these vehicles at any one time. The ground system cannot isolate the two (2) turned-around spacecraft emissions since they will be at the same frequency. This situation will not hold only after transposition when the S-IV B and LEM may have different radial velocity components. Consequently, it is necessary that the S-band transponder aboard one of these two (2) vehicles be off when it is desirable to communicate with the other vehicle. The two (2) transmitted uplink frequencies are very stable, being derived from a time standard having a stability of approximately 5 parts in 10^{11}. This high stability in the transmitted frequency gives rise to the accurate doppler measurements of the Unified S-Band System.

The modulation waveform applied to the uplink carrier consists of basically two (2) types. These are the modulated subcarriers and the binary ranging code. There are two (2) uplink subcarrier frequencies obtained from two (2) independent subcarrier oscillators. The voice subcarrier is derived from a 30 kilocycle oscillator that is frequency modulated with voice from 100 cps to 3 kilocycles per second. Another subcarrier oscillator operating at 70 kc is modulated with digital commands. This modulation consists of a 1 kc sync tone and a 2 kc sine wave bi-phase modulated at an information rate of 200 cps. The other type of uplink modulation is the binary ranging code. This ranging code is derived from the ranging subsystem, the details of which are beyond the scope of this paper. Basically, four (4) digital code generators are driven at twice the receiver clock frequency which is about 498 kc. Each code generator is set up to repeat its code at some multiple of the basic 1 microsecond (approximate) bit period. This multiple is a relative prime number in that the lengths of the code generators have no common denominator. The outputs of the four (4) code generators are combined with twice the receiver clock frequency (approximately 1 Mc) through digital logic to form a "maximum linear code." This code contains a total number of bits before repetition equal to the multiple of the lengths of the individual code generators and clock component. It is possible to generate two (2) different codes in the USB System. The short code, used primarily during launch and earth orbit, is approximately 70 milliseconds in length. A longer code can be generated for lunar distances and has a code length of about 5.4 seconds. The unambiguous range of the long code allows ranging at distances considerably larger than lunar distances.

Figure 8 illustrates in principle the basic USB uplink spectrum. The generated ranging code is summed with the voice and up-data subcarriers. It is very important to adjust the subcarrier and ranging code phase deviations in order that sufficient power is contained in each of these components. The amount of power in each of these three (3) components is dependent upon the value of a Bessel function whose argument is the phase deviation of the components under consideration. In addition, it is extremely important to have a strong carrier component in the uplink spectrum in order that the phase-lock loop in the spacecraft transponder can adequately lock to this frequency. The threshold levels of the various detectors in the transponder determines the amount of power required in each component of the uplink spectrum which in turn determines the phase deviation required for each compenent in the modulation process. Since the limiting case of the binary ranging spectrum is a square wave, its Fourier Transform results in a sin x/x distribution in the frequency domain, that is symmetrical about the carrier. The first nulls occur at the inverse of a bit period or approximately 1 Mc. Figure 8 also illustrates the S-Band two-sided spectrum where the voice subcarrier components are located ±30 kc about the carrier and the up-data subcarrier components are located ±70 kc about the central carrier.

THE VOICE, UPDATA AND BINARY RANGING CODE ARE COMBINED AND THE RESULTANT PHASE MODULATES THE S-BAND CARRIER TO FORM THE FOLLOWING SPECTRUM FOR UPLINK TRANSMISSION

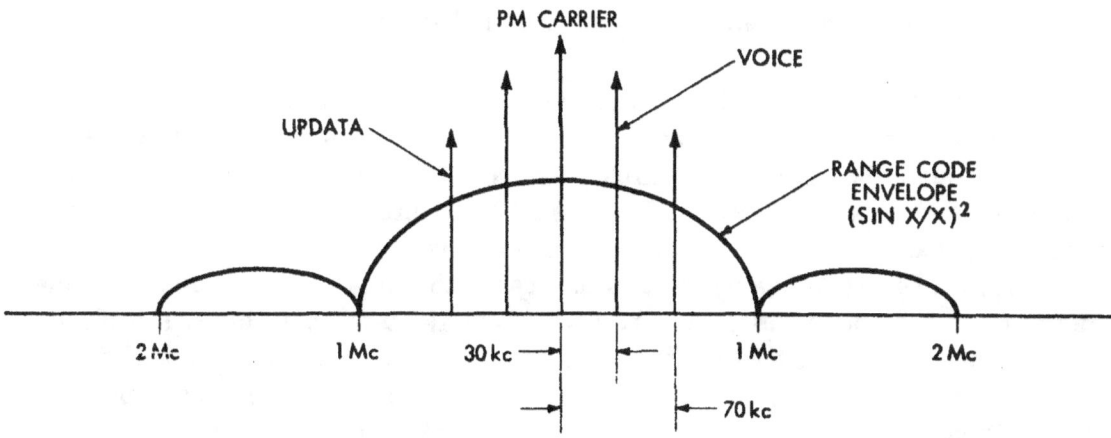

Figure 8. Uplink Spectrum

Figure 9 is a very basic block diagram of the spacecraft USB transponder. The spacecraft is equipped with two (2) S-band antennas. An omnidirectional antenna is used in all near-earth phases through transposition. After transposition is complete, the spacecraft high-gain array is oriented so that it can communicate with the earth ground station. This array has three (3) possible modes of operation resulting in gains of 14, 22, or 28 db, respectively. There is no preamplifier in the front end of the transponder. Thus, the received signal is fed directly to a mixer having a noise figure of about 11 db. The S-band spectrum is converted to an i.f. frequency in the PM receiver. A narrow-band carrier tracking phase-lock loop having a loop bandwidth of about 1 kc locks to the uplink carrier and continues to track it as the frequency changes with spacecraft velocity. The spectrum is fed to a wideband phase detector where the ranging code is recovered at video for later recombination with the downlink modulation. The uplink voice is extracted and monitored by the astronauts while the digital commands are recovered and fed to the on-board computer where they can be converted into useable instructions for the on-board systems.

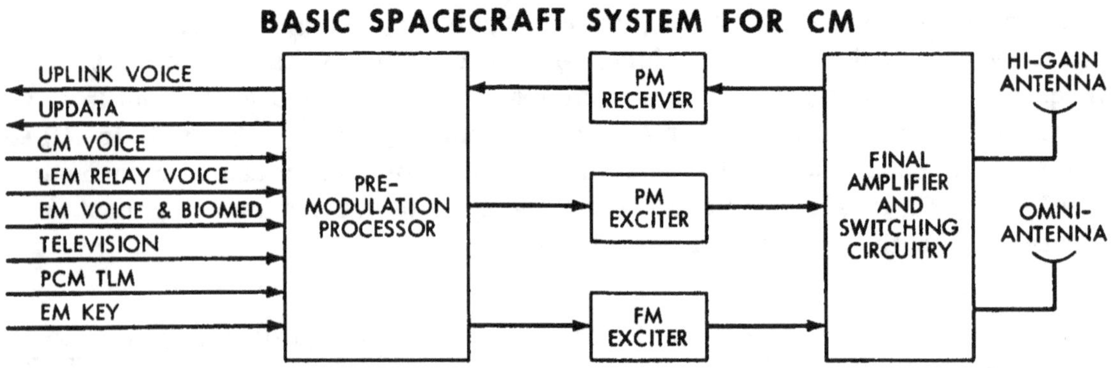

BASIC SPACECRAFT SYSTEM FOR CM

Figure 9. Basic Spacecraft System for Command Module

The universal premodulation processor is a common acceptance and distribution point for all information that has to be transmitted from the spacecraft to ground via the USB downlink. There are many modes in the S-band downlink, and it is necessary to have one (1) common piece of equipment where a selection can be made as to the type of information to be transmitted. For example, there are times when it is desirable to transmit only voice and telemetry from the spacecraft. Another time may demand that voice, telemetry, and ranging be transmitted to the ground station. It should be noticed that the requirement for ranging in the downlink is a function of whether ranging was transmitted to the spacecraft. Also, there may be times when it is necessary to transmit both the PM mode and the FM mode from the command module simultaneously. Since the downlink is a complex spectrum similar to the uplink, it is necessary to carefully select the proper amount of modulation for each type of information transmitted in order to optimize the power in each component. The phase deviation for each information component for each mode can be made in the premodulation processor. It is necessary to adjust the phase deviations for each mode in order to share the transmitter power between components in such a way that the various detection process in the ground station is optimized from a signal-to-noise ratio aspect. Figure 9 illustrates the various inputs to the premodulation processor from the on-board systems. In essence, the premodulation processor is a large switching and processing network that selects the information to be transmitted and adjusts the amount of modulation for each information component.

The uplink carrier frequency received by the USB transponder in the spacecraft is translated by a factor of 240/221 before modulation by the downlink information. This new frequency, which is derived from the voltage-controlled oscillator in the transponder carrier tracking loop, is modulated by the downlink video spectrum and fed to the final power amplifier which is a 20 watt traveling wave tube. In addition to the PM exciter, there is an FM exciter in the transponder which is necessary for the FM mode. The FM mode is used primarily at lunar distances when it is necessary to transmit television. It is possible to transmit the PM and FM modes simultaneously from the CM, but not from the LEM. The ground station will have the capability to receive and demodulate both of these modes simultaneously. The LEM can transmit only PM or FM, but not both simultaneously. Since the LEM transmit frequency is the same for either FM or PM, it is desirable that the LEM uplink not contain ranging information during the times when it is desirable to send FM television from the spacecraft. The capability exists to transmit telemetry and voice in the PM and FM mode for both the LEM and CM.

Figure 10 lists some of the salient features of the Unified S-Band spacecraft transponder. The TWT power amplifier has the capability to put out 20 watts of CW power. However, there is another mode whereby the output power can be reduced to 5 watts. The TWT feeds either one (1) of two (2) antennas through a switching network. As previously mentioned, the omnidirectional antenna is used from launch through transposition. Transposition is the phase of the mission that occurs after injection into a translunar flight trajectory.

SPACECRAFT USB CHARACTERISTICS

1. POWER AMPLIFIER - TWT
   POWER OUTPUT - 20, 5 WATTS

2. ANTENNAS
   OMNIDIRECTIONAL (NEAR EARTH THRU TRANSPOSITION)
   HIGH GAIN (28 db) (AFTER TRANSPOSITION AND LUNAR)

3. CARRIER FREQUENCIES
   2287.5 Mc - COMMAND MODULE PM CARRIER
   2272.5 Mc - COMMAND MODULE FM CARRIER
   2282.5 Mc - LEM OR SIV-B PM CARRIER
   2277.5 Mc - SIV-B FM TELEMETRY CARRIER
   2282.5 Mc - CAN ALSO BE USED IN FM MODE FROM LEM
   NOTE: ALL PM SPACECRAFT CARRIERS ARE $\frac{240}{221}$ x GROUND PM CARRIER

4. VOICE SUBCARRIER - 1.25 Mc
   MODULATION - 300 TO 3000 cps VOICE + 7 CHANNELS OF BIOMEDICAL DATA

5. TELEMETRY SUBCARRIER - 1.024 Mc
   BI-PHASE MODULATED AT
   51.2 kbps - HIGH DATA RATE
   1.6 kbps - LOW DATA RATE

6. FM TELEVISION
   500 kc MODULATION
   320 LINES
   10 FRAMES/SEC

7. EMERGENCY KEY
   512 kc SUBCARRIER MANUALLY KEYED AND MODULATED DIRECTLY ON CARRIER

Figure 10. Spacecraft Unified S-Band Characteristics

During this maneuver, the S-IVB is separated completely from the CM/LEM combination. In addition, the LEM separates temporarily from the CM, turns around, and relocks with the CM. It is expected that this maneuver will be over at approximately 12,000 miles from earth. During this time, two-way communications will be established using the omnidirectional antenna on the spacecraft and the 30-foot ground antenna. This phase of the mission is perhaps the most critical as far as reliable communications are concerned. This is due to the fact that high-bit-rate telemetry from the spacecraft is a requirement during this phase of the mission and in certain regions the omnidirectional antenna gives negative gain. This telemetry link is, in general, the most critical link during any phase of the mission from a signal-to-noise ratio required for detection viewpoint. After transposition, the high-gain array is used as the primary spacecraft antenna in either the 14 db, 22 db, or 28 db gain mode. At this time a strong two-way link can be established.

Figure 10 lists the S-band frequencies that are available from the transponders on the CM, LEM, and Saturn S-IVB. The CM PM carrier is approximately 2287.5 Mc, being a function of the uplink PM carrier. As mentioned earlier, the uplink carrier is multiplied by a factor of 240/221 to produce the downlink carrier. Since the two-way Doppler frequency can be as high as 120 kc, the nominal 2287.5 Mc vary by ±120 kc. The FM carrier is generated in the spacecraft and its nominal frequency is 2272.5 Mc. These two (2) carriers can be transmitted simultaneously or separately depending upon the mode of operation. In general, television is not a requirement until the lunar phase, thus, two (2) links from the CM could be expected at lunar distances.

The other uplink carrier previously mentioned was 2101.8 Mc. This carrier can be utilized for communications with either the LEM or the Saturn S-IVB. The S-band transponder on these vehicles also multiplies the uplink carrier by 240/221 to give a downlink PM carrier of 2282.5 Mc. Since it is important that only one (1) of these two (2) vehicles utilize this carrier at a time, it is necessary to time share this frequency between the LEM and S-IVB. This time sharing depends upon the phase of the mission and operational considerations. Since the LEM has only one (1) transmit frequency, it is necessary to utilize this carrier for both the FM and PM modes; however, these modes cannot be transmitted simultaneously from this vehicle. The S-IVB, having no requirement for television, has only a PM mode capability primarily for the purpose of tracking. In addition, a PCM S-band telemetry transmitter is utilized on the S-IVB and has the capability to transmit 70 kilobits of information to the ground station at 2277.5 Mc. Thus, it is possible to have as many as four (4) downlink frequencies being received and demodulated by a dual ground station simultaneously.

Figure 10 lists the various subcarriers that can be transmitted from either the LEM or CM. A 1.25 Mc voice subcarrier is modulated with voice from 300 cps to 3000 cps and 7 biomedical channels that occupy the frequency range from 3000 cps to about 12 kc. The other basic subcarrier generated in either of the spacecraft is the 1.024 Mc telemetry subcarrier. This subcarrier frequency is bi-phase modulated at 51.2 kilobits per second in the high-bit-rate mode and 1.6 kilobits per second in the low-bit-rate mode. Voice and telemetry are the two (2) basic types of spacecraft information that must be transmitted during most of the mission. In addition, television is required at lunar distances. This TV system operates with a modulating frequency of 500 kcps and has a 320 line capability resulting in pictures at the rate of 10 frames per second.

There are several back-up modes available from either the LEM or the CM. The emergency key mode consists of manually keying a 512 kc sub-carrier which is then modulated directly onto the S-band carrier. An emergency voice mode consists of modulating the S-band carrier directly with voice from 300 to 3000 cps. These modes are used only if a major failure occurs in the spacecraft transponder.

Figure 11 represents typical spectrums of the possible modes that can be transmitted from the spacecraft. The top spectrum indicates the basic PM mode originating from either the LEM or CM. It should be noticed that the subcarriers are in the first sidelobe of the (sin x/x) ranging spectrum and occur reasonably close to the first null. A strong carrier component is required in this mode in order that the ground receiver can lock to the spacecraft frequency. The FM mode, as shown, also contains the two (2) subcarriers in addition to the TV spectrum. As shown, the carrier component disappears and the FM mode is strictly a data mode; not a tracking mode. The ranging spectrum in the downlink results from the turned-around uplink ranging code on the downlink carrier. The detail of the modulation on the subcarrier is not shown due to the lack of space on the figure. It should be noted that voice does not originate from the S-IVB; thus, the voice subcarrier would be missing in the PM spectrum of the S-IVB. The bottom two (2) spectrums are those of the emergency voice and emergency key modes. The PM mode is the basic S-band tracking and data mode in that most of the normal two-way communications can be accomplished except for transmission of TV from the spacecraft.

Figure 12 lists the basic functions of the USB spacecraft transponder. This list is by no means complete, but the basic operations performed are covered. The transponder has a narrow-band phase-lock loop operating at an intermediate frequency. This loop locks to the carrier component in the uplink spectrum and continues to track it as the frequency changes due to the change in the radial component of the spacecraft velocity. The loop has been designed

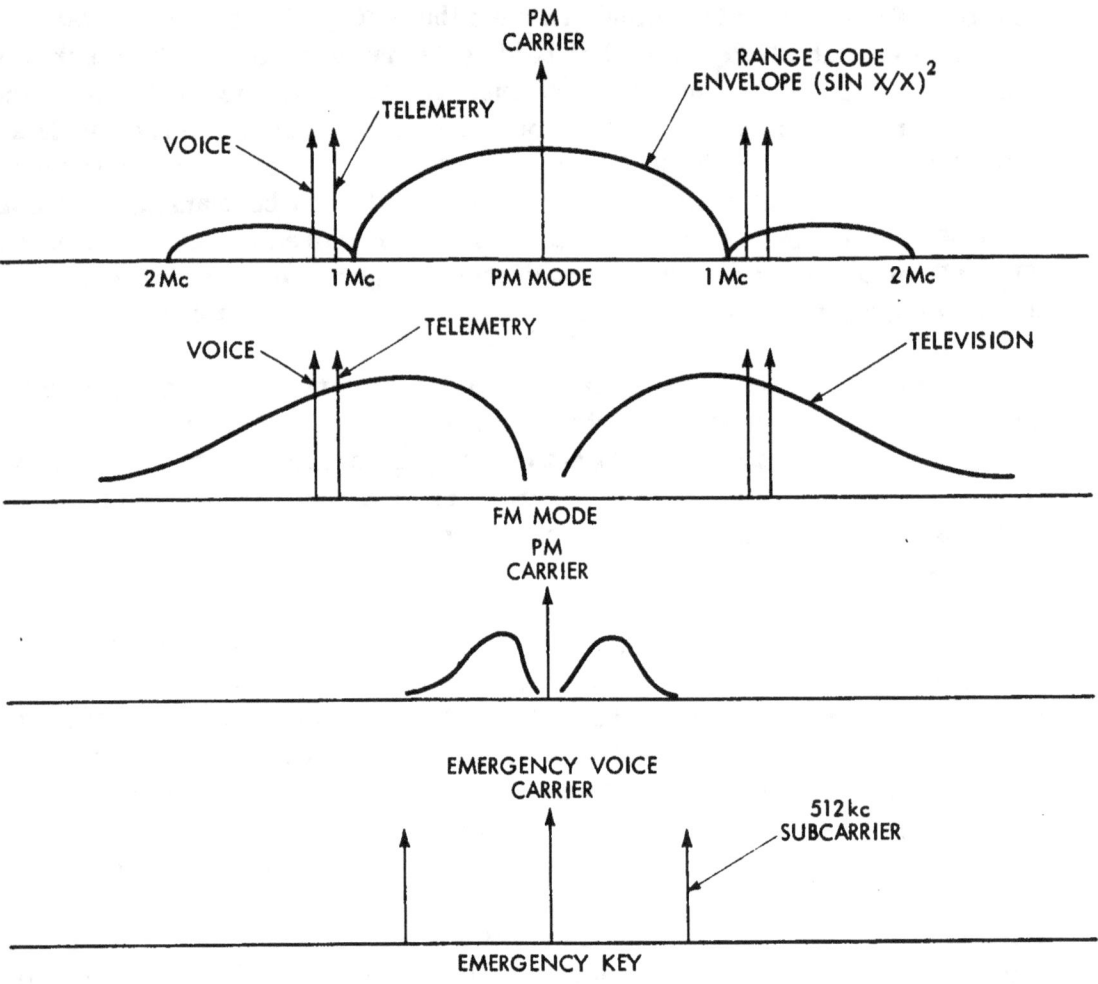

Figure 11. Downlink Spectrums

so that it can track anticipated changes in the one-way Doppler frequency. In addition, the voice and up-data subcarriers are filtered out, demodulated, and the voice is monitored by the astronauts while the up-data commands are fed to the on-board computer. The binary ranging code is detected and converted to video for remodulation of the downlink carrier. It is important that any phase delays added in the transponder be known precisely. This is due to the fact that the ranging measurement made in the ground receiver relies upon a completely coherent two-way link with the spacecraft. Any additional time delays

FUNCTIONS OF S/C USB TRANSPONDER

1. LOCK ON TO UPLINK CARRIER AND TRACK IT AS THE FREQUENCY CHANGES DUE TO DOPPLER EFFECT.
2. EXTRACT THE UPLINK SUBCARRIER MODULATION TO PRODUCE VOICE AND COMMAND INFORMATION.
3. DETECT BINARY RANGING SIGNAL IN A WIDE BAND PHASE DETECTOR AND TRANSLATE TO A VIDEO SIGNAL.
4. TRANSLATE RECEIVED FREQUENCY BY A FACTOR OF 240/221.
5. COMBINE TELEMETRY AND VOICE SUBCARRIERS WITH RANGING SIGNAL AND THE RESULTANT MODULATES THE PM DOWNLINK CARRIER.
6. COMBINE TELEMETRY AND VOICE SUBCARRIERS WITH TELEVISION AND THE RESULTANT MODULATES THE FM CARRIER.

Figure 12. Functions of Spacecraft Unified S-Band Transponder

added in the transponder would appear to the ground station as an increase in spacecraft range; thus, if these delays are not known exactly, an error will result in the range measurement.

The downlink PM carrier derived in the spacecraft is 240/221 times the received uplink carrier. This frequency is derived from the voltage controlled oscillator in the transponder phase-lock loop. The modulated voice and telemetry subcarriers are summed with the turned-around binary ranging code and the resultant waveform modulates the PM downlink carrier. For the FM mode, the subcarriers are summed with the TV video signal and this resultant modulates the FM carrier. Here is an example of the difficulty involved in attempting to optimize the modulation indices of the various information components. In order to optimize the ground detection processes, it is necessary to adjust the indices for each mode of operation. The premodulation processor serves the function of readjustment in modulation indices.

Figure 13 presents the basic functions of the USB ground station equipment. The first important function served is that of early acquisition as the spacecraft appears on the radio horizon. Early angle information is sufficient to allow the main beam of the 30-foot or 85-foot antenna to be positioned on the spacecraft. In a short time the difference pattern developed by the feed system for the X and Y axes feeds an error to the servo control system that in turn places the electrical axis of the large antenna on the spacecraft. The sum pattern developed by the feed system is fed to the sum channel receiver. The PM channel receiver utilizes a very narrow-band phase-lock loop in the front end for the purpose of locking to the downlink carrier frequency. This loop continues to track this frequency throughout the pass of the spacecraft. The Doppler frequency shift associated with radial spacecraft velocity is derived from the voltage-controlled oscillator in this loop by comparing its frequency with the transmitter VCO frequency. Since the received frequency has a definite known phase

FUNCTIONS OF USB GROUND STATION

1. THROUGH THE AID OF THE ACQUISITION ANTENNA, ACQUIRE THE SPACECRAFT IN ANGLE.
2. USE POINTING INFORMATION TO POSITION 30 FOOT OR 85 FOOT ANTENNA BEAM ON SPACECRAFT.
3. CASSEGRAIN FEED SYSTEM DEVELOPS A SUM AND A DIFFERENCE PATTERN ON RECEIVED SIGNAL.

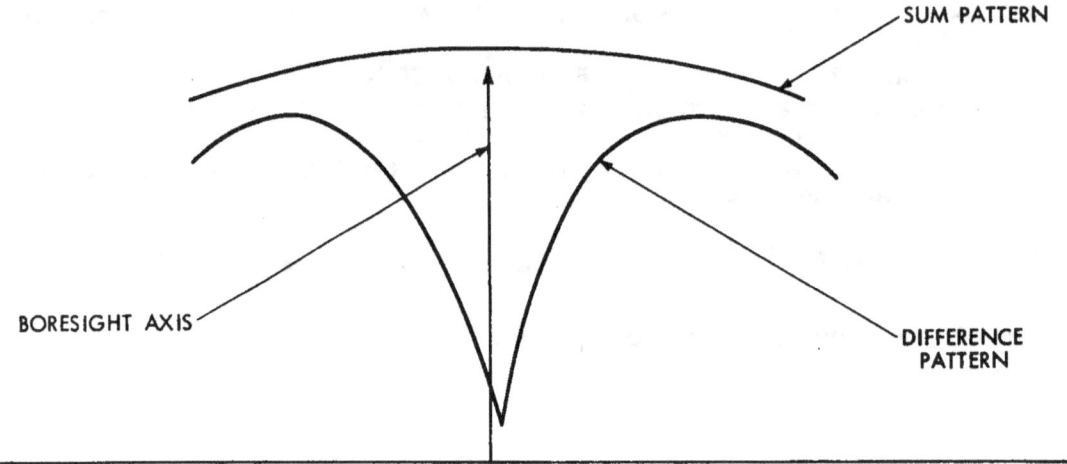

4. X AND Y AXIS DIFFERENCE PATTERN SIGNAL USED TO KEEP ANTENNA BORESIGHT ON SPACECRAFT.
5. THE SUM PATTERN IS USED FOR THE PROCESSING OF ALL INFORMATION IN THE GROUND RECEIVER.
6. THE CARRIER TRACKING PHASE LOCK LOOP LOCKS TO AND TRACKS DOWNLINK PM CARRIER AS IT CHANGES DUE TO DOPPLER SHIFT.
7. THE DOPPLER FREQUENCY IS DERIVED FROM THE VOLTAGE CONTROLLED OSCILLATOR IN THE CARRIER TRACKING P.P.L.

THE RECEIVED BINARY RANGE CODE IS COMPARED WITH THE RECEIVER RANGE CODE AND THROUGH AUTO CORRELATION THE SPACECRAFT RANGE IS DETERMINED.

THE PM CARRIER IS CONVERTED TO A 10 Mc I.F.

THE FM CARRIER IS CONVERTED TO A 50 Mc I.F.

ALL DOWNLINK INFORMATION IS DEMODULATED IN THE SIGNAL DATA DEMODULATORS AND SENT TO THE PROPER DATA PROCESSING EQUIPMENT.

Figure 13. Functions of Unified S-Band Ground Station

relationship (coherent) to the transmit frequency, Doppler frequency can be measured very accurately.

The ranging code in the received downlink spectrum is sent to the range clock receiver and code clock transfer loop. Here, the received ranging code is compared with the receiver ranging code through an autocorrelation process. The time elapse required for these two (2) codes to completely correlate represents the two-way propagation time to the spacecraft, which can be converted into range to the spacecraft. This autocorrelation process is similar to autocorrelating a pulse in a radar application. However, the time delay, τ, ordinarily introduced with delay lines is accomplished in the ranging system by digitally shifting the individual code components.

The received S-band signals are converted in the sum receiver to a 10 Mc PM output and a 50 Mc FM output. Each output feeds a separate set of signal data demodulators. Within each set of demodulators, the telemetry and voice subcarriers are filtered out and demodulated. The telemetry bit stream is processed to the telemetry computer via the PCM processor, and the voice is either monitored by the site personnel or relayed to the MCC. In addition, the TV from the FM mode is demodulated and fed to a TV monitor and tape recorder at the USB station.

A carrier phase demodulator follows the 10 Mc output of the sum receiver and consists of a narrow-band phase lock loop that tracks the 10 Mc carrier component. A 1.25 Mc filter for extracting voice and a 1.024 Mc filter for extracting the telemetry subcarrier are placed at the output of the phase detector in this loop. Also tied to the phase detector output is a low-pass filter for extracting the emergency voice and a filter to extract the emergency key component. The telemetry and voice demodulators utilize a phase-lock loop detector. A wideband modulation tracking phase-lock loop follows the 50 Mc output of the sum receiver. This loop demodulates the TV spectrum and the output of the loop filter is fed to the TV monitor. In addition, this loop feeds a telemetry and voice demodulator that is identical to that used in the 10 Mc channel.

Figure 14 is a very basic description of one method that might be utilized for achieving two-way frequency lock between the ground station and spacecraft. The acquisition procedure is being finalized at this writing and Figure 14 is included only to suggest to the reader one possible acquisition procedure.

Figure 15 represents just a few of the system highlights of the USB System. It is important to note that with the USB System two (2) spacecraft can be communicated with simultaneously if they are both in the beamwidth of one (1) antenna. However, with the JPL back-up 85-foot stations, it is possible for two (2) spacecraft to be tracked independently and simultaneously by allowing the JPL site and the MSFN 85-foot site to each track a separate vehicle at lunar distance. When one (1) site tracks two (2) vehicles, it should be noted that angle information cannot be developed separately on both vehicles. However, independent range and Doppler measurements can be made for each spacecraft in this situation. This implies that full two-way communications is carried on with both vehicles and a dual ground station. There are times when a dual ground station (primarily 30-foot in this case) can receive from three (3) spacecraft simultaneously. For example, the CM and LEM could each be transmitting in the PM mode and the S-IVB could be transmitting the 2277.5 Mc telemetry link, resulting in three (3) different links that have to be demodulated simultaneously.

POSSIBLE METHOD OF ACHIEVING 2-WAY LOCK BETWEEN SPACECRAFT AND GROUND STATION

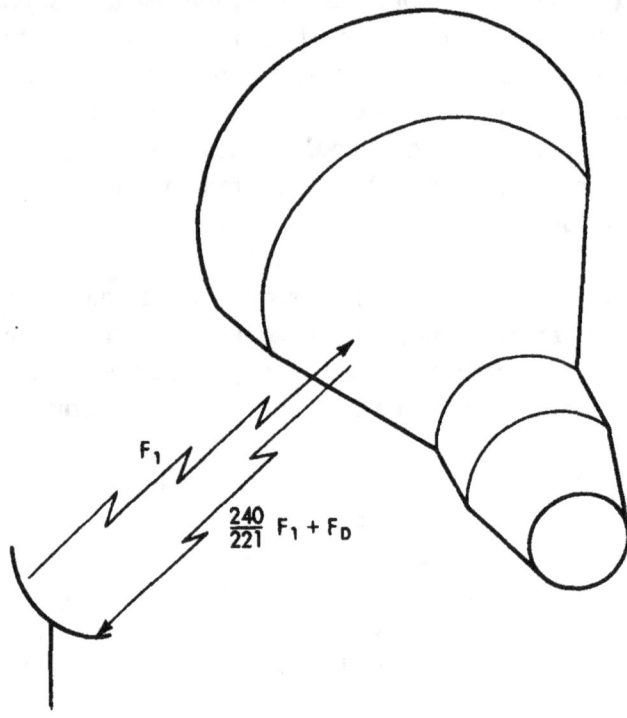

1. CARRIER WITHOUT RANGE CODE TRANSMITTED TO SPACECRAFT.
2. NARROW BAND PHASE LOCK LOOP IN SPACECRAFT TRANSPONDER LOCKS TO GROUND CARRIER FREQUENCY.
3. AFTER TRANSPONDER LOCKS TO GROUND FREQUENCY; THE SPACECRAFT TRANSMITS THE FREQUENCY $240/221\ F_1$.
4. MAIN CARRIER TRACKING LOOP IN GROUND STATION LOCKS TO S/C FREQUENCY ACHIEVING 2-WAY COHERENT LOCK.
5. UPON COMPLETION OF 2-WAY LOCK, THE UPLINK AND DOWNLINK CAN BE FULLY MODULATED.

Figure 14. Possible Method of Achieving Two-Way Lock Between Spacecraft and Ground Station

SYSTEM HIGHLIGHTS

1. SIMULTANEOUSLY COMMUNICATE WITH 3 SPACECRAFT WHEN ALL ARE IN 1 ANTENNA BEAMWIDTH.
2. FLIGHT CONTROLLERS HAVE BEEN ELIMINATED FROM MANY OF THE REMOTE SITES.
3. DIGITAL COMMANDS CAN BE GENERATED BY COMMAND COMPUTER AT THE REMOTE SITES.
4. ANGLE RESOLUTION 0.025°.
5. DOPPLER RESOLUTION 0.1 METERS/SEC.
6. RANGE RESOLUTION 1.5 METERS.

Figure 15. System Highlights

The Apollo program is unique in that flight controllers have been eliminated from many of the remote sites. This has resulted in the requirement that most of the data processed at the site be sent to either GSFC or the MCC at Houston. Since this data consists of tracking data, telemetry, and spacecraft voice, it is necessary to have high quality communications circuits between the remote sites and the data user. In addition to teletype, high-speed (2400 bps) circuits will be available into and out of most of the sites. Thus, the NASCOM communications network is an integral part of the Apollo MSFN Network program. This aspect is important since the information received by the flight controllers at the MCC is used to make decisions that may require digital commands or voice to be sent to the spacecraft.

The Apollo digital command system is unique. Each site is equipped with a 642B Univac computer that can be loaded with commands from the MCC during the pre-pass. As the spacecraft appears on the radio horizon, commands that have been loaded in memory can be called up for transmission to the spacecraft at a designated time. In addition, there is also a real time mode where the MCC can send commands directly to the S/C. For this case, the command computer acts as a real time distribution center for the commands.

The USB System will give an angular accuracy of 0.025 degrees. This is the worst case analysis and includes all possible errors ordinarily considered. The velocity resolution of the USB is excellent resulting in 0.1 meters per second. In addition, the expected range resolution is about 1.5 meters.

SUMMARY

It is realized that this paper is only a bird's-eye view of the Apollo USB system. The Apollo mission called for a unique telecommunications system with a capability to provide primary tracking and communications. This has been designed into the USB by combining the JPL ranging system with data subcarriers for both the ground-to-spacecraft and spacecraft-to-ground links. Utilization of large ground antennas and optimization of spacecraft and ground modulation and demodulation processes will insure a reliable, coherent two-way communications link for Project Apollo.